Henrietta Lacks

The Legend of Henrietta Lacks

By Naven Johnson

Copyright © 2018 by Naven Johnson

All rights reserved. This book or any portion thereof may not be reproduced or used in any manner whatsoever without the express written permission of the publisher except for the use of brief quotations in a book review.

Printed in the United States of America

First Printing, 2018

Table of Contents

Who Was Henrietta Lacks 4

What are Immortal Cells 9

How Lacks' Immortal Cells Were Utilized 13

Advancements in Cancer Research 19

Cell Discoveries and Complications 23

Family Retribution and Recognition 28

Where to Learn More 32

Who Was Henrietta Lacks

On a bright day of August 1st, in the year of 1920, Eliza and Johnny Pleasant brought forth a girl Loretta Pleasant, who's name was later on changed to Henrietta Lacks for reasons unknown to the family. As she grew up, she was given the nickname Hennie.

At four years of age her birth mother died of birth complications from her tenth child. Following the hardship of taking care of the children solely, after the demise of his wife, Johnny moved to Clover, Virginia. He then gave guardianship of his children to his folks. Lacks moved in with her grandfather Tommy Lacks, in a two-story log cabin initially owned by her great grandfather and great uncle (it once served as slaves' quarter on the farm). She shared a room with David 'Day 'Lacks, nine years old at the time. He was her cousin, who would later be her husband, and had been there since 1905.

From a tender age, Lacks became a tobacco farmer just like most of her relatives in Clover. At only 14 years she gave birth to a son, Lawrence Lacks, and five years later in 1939 she bore a daughter, Elsie Lacks. Both sired by Day Lacks. Elsie had developmental disabilities which made her viewed as 'different', or 'deaf and dumb.'

On the memorable day of 10th April 1941, Day and Hennie got married in Halifax County, Virginia. They later moved to Maryland, leaving their tobacco farm following much convincing from Fred Garrett, their cousin. Shortly after, Day started working at Bethlehem Steel in Sparrow's Point, while Fred was called to duty to fight in World War II. From grant savings of his cousin, Day bought a house at 713 New Pittsburgh Avenue in Turner Station. It was one of the oldest, and mostly African American communities in Baltimore County, which is the present Dundalk, Maryland.

It is while at Maryland that they got three more children; David 'Sonny' Lacks in 1947, Deborah Lacks-Pullum 1949-2009 and Zakariyya Bari Abdul Rahman born Joseph Lacks in November 1950. Her last born, Joseph, was born at Johns Hopkins Hospital, four and a half months before Hennies' diagnosis with cervical cancer. At the time, Elsie lacks was admitted to the Hospital for the Negro Insane, which was later renamed as Crownsville Hospital Center. She succumbed to her cancer in 1955. Johns Hopkins was the only hospital in the area that treated black patients. On the 29th of January 1951 Lack went for a checkup, since she felt a knot in her womb. This was something she had already discussed with her cousin, and dismissed it as pregnancy, although after birth she had a severe hemorrhage. Her doctor figured it might be syphilis but she tested negative for syphilis, and then she was once again referred back to Johns Hopkins.

It was at Hopkins that Dr. Howard W. Jones took a cervix mass for lab testing, and Lack was informed of malignant epidermoid carcinoma presence in the uterus, a misdiagnosis as physicians would later discover in 1970.

She had been suffering from adenocarcinoma. The diagnosis was a mistake, however, treatment would not have differed. She was treated with radium tube inserts and discharged with instructions to return for X-ray treatment. During her visits, two samples were taken from her subconsciously; one which was a tough or healthy tissue, and the other cancerous. George Otto Gey, who was an excellent physician and cancer researcher, took the cancerous sample. It was later referred to as HeLa immortal cell line when it was used in biomedical research. At an undisclosed family cemetery, Hennie was buried when she passed away in Halifax, Virginia. The Lackstown name was adopted for Clover Land, formerly owned by white slave owners before the civil war. The land was later taken up by the Lacks family who were descendants of African slaves, and the white owners.

While Lacks burial site is a mystery, the family believes it lies a few meters from her mother's grave, which had been the only one on the grounds with a marking or tombstone. In 2010, Roland Pattillo of Morehouse School of Medicine and George Gey, a colleague and Lacks family friend, donated a headstone for Lacks. Consequently, the family raised money for Elsie Lacks' as well, and they were dedicated on the same day. Henrietta's headstone is book-shaped containing epitaph written by her grandchildren.

What are Immortal Cells

An immortalized cell line primarily refers to a growing population of cells from a multicellular organism which, under normal circumstances, doesn't proliferate indefinitely. Yet due to mutation, it had evaded the normal cellular senescence, and kept on undergoing modifications. Mutations for immortality can occur naturally, or induced for experimental purposes. Immortal cells are important in biochemistry, biotechnology and cell biology of multicellular research.

An immortalized cell is not a stem cell which can also divide indefinitely, but form a regular part of the multicellular organism development.

Immortal cell lines are either normal cell lines derived from a stem cell, or the intro equivalent of cancerous cells. Cancer occurs when a somatic cell which cannot divide, undergoes mutations which cause a deregulation of cell cycle controls, leading to uncontrolled proliferation. Immortalized cell lines have undergone similar modifications allowing a cell type not to split. The origins of some immortal cell lines are with naturally occurring cancer.

Immortalized cell lines can be a simplified model for more complex biological systems, as in the analysis of the chemical nature and cell biology of mammalian (including human) cells. The immortality of the immortal cell line becomes the striking advantage, as a cell can be grown indefinitely in culture, which then simplifies cell biology analysis, which would have been challenging and limited.

Cloning of these cell lines has given rise to a clonal populace of cells which can be propagated indefinitely, and gives room for repeated scientific experiments, and analysis on the genetically identical cells. The alternative way of analyzing primary cells from multiple tissue donors do not have this advantage as of the latter.

In biotechnology, immortalized cell lines are a viable way of growing cells similar to those in a multicellular organism which is in vitro. They are used for testing toxicity of compounds, and production of eukaryotic proteins among others.

They often originate from a popular tissue type that has undergone several mutations to become immortal; this can alter the cell biology, and hence must be considered in any analysis. These cell lines are dynamic as they change genetically over many passages, resulting in differences in phenotypic state, and experimental outcomes subject to when, or with what cell strain is experimented with.

Most cell lines in biomedical research have been contaminated, and overgrown by other more aggressive cells. For instance, what was thought to be thyroid lines were really melanoma cells; the alleged prostate tissue turned out to be bladder cancer, while the common uterine cultures in the real sense were breast cancer.

How Lacks' Immortal Cells Were Utilized

George Otto Gey had propagated Lacks' cells before she died which was the first successful propagation in vitro, an achievement with profound future benefit to medical research. Gey freely donates these cells to any scientist requesting them for research. He had obtained them neither with Lacks' consent, or her families. At the time it was neither required nor customarily sought. The cells have been commercialized but never in the patented original form.

There is no requirement presently, or in the old days to inform patients, or relatives of any material obtained during surgery, diagnosis, or therapy since it belongs to the physician, according to the ethical approval and patient consent in the UK. The scenario surrounding

Lacks' tissue removal was tabled in the Supreme Court of California as a case of Moore vs. Governors of the University of California. The court ruled that a person's discarded tissue and cells are not his or her property, and can be commercialized.

Initially, the cell line was to be 'Helen Lane' or 'Helen Larson' to conceal the fact that Lacks' cells were taken without consent, but despite the attempts the press still used the real names. These cells are cancerous as they were obtained from a biopsy taken from a visible lesion on the cervix as part of Lacks' cancer diagnosis. The debate however still goes on in regards to the classification.

HeLa cells are immortal since they can divide an unlimited number of times in a lab cell culture as long as basic cell survival conditions are present. There are many types of HeLa cells as they continue mutating in cell cultures, but all HeLa cells are descended from the similar tumor cell from Lacks. The propagated number of HeLa cells by far exceeds the cells which were once present in Henrietta Lacks' body as a whole.

Jonas Salk used the HeLa cell to test the first polio vaccine in 1950. It was observed to be easily infected by poliomyelitis causing cell death, which made it ideal for polio vaccine testing since results were visible. Nevertheless, a significant number of HeLa cells were required, prompting the National Foundation for Infantile Paralysis (NFIP) to find a facility for mass HeLa cells production. In 1953, at Tuskegee University, a cell culture factory was founded to supply Salk and other labs with HeLa cells. Within one year, Salk's vaccine was ready for human trial.

Theodore Puck, and Phillip I Marcus at the University of Colorado, successfully cloned HeLa cells. These were the first cloned human cells, and this was achieved in 1955. Since then, the cells have been used for research into cancer, AIDS, effects of radiation and toxic substances, gene mapping and many others. According to Rebecca Skloot's book of 2009, over 60000 research articles on

HeLa cell had been published and the numbers rapidly on an upward trend at a rate of 300 research papers monthly. In expounding, she gave the classic example of Chester M. Southam's cancer injection experiments on patients and also those of prison inmates from the Ohio State Penitentiary.

HeLa cells have widely been used in testing how parvovirus infects, human, HeLa, dog, and cat cells. They have also been used to study Oropouche virus (OROV), which causes disruption of cell culture where cells disintegrate after infection, causing viral induction of apoptosis. Similarly, they are used to examine the expression of papillomavirus E2 and apoptosis. In the study of canine distemper virus ability to induce apoptosis in cancer cell lines, it's helpful in developing the treatment for tumor cells resistance to chemotherapy and radiation.

In some recent cancer studies, HeLa cells are used in sex steroid hormones e.g. the estradiol, estrogen, its receptors, estrogen-like compounds such as quercetin, and its reducing properties on cancer cells. Related studies on the effects of flavonoids and antioxidants with estradiol on cancer cell proliferation with the use of the same were conducted. Their applications extend to the investigation of phytochemical compounds, and working on the anticancer activity of the ethanolic extract of mango peel (EEMP). It was found to contain various phenolic compounds in addition to activating the death of human cervical HeLa malignant cells through apoptosis, which suggests that EEMP can prevent cervical and other cancers.

In 2011, HeLa cells were used in tests of a new heptamethine dye called IR-808, and other analogs which are currently explored for their unique applications in medical diagnostics, development of theranostic, personalized treatment as per the cancer patient with either PDT, irradiation, or it can also be co-administered with other drugs. In research involving fullerenes, HeLa cells have also been used to induce apoptosis as a photodynamic therapy, and in -Vitro cancer research which utilizes the cell lines. At great lengths, HeLa cells have also been used to show cancer markers in RNA, and establishment of an RNAi-based Interference of Specific cancer cells and their system identification.

Advancements in Cancer Research

The HeLa cells proliferate quite fast in comparison to other cancer cells used in medical research. Just like other cancerous cells, HeLa's exhibit one version of Telomerase during cell division, preventing incremental shortening of the latter, which results in aging and eventual cell death. Through this, the HeLa cells exceed the Hayflick Limit. This limit is based on the number of times a cell can divide before it turns senescent.

The HeLa genome created by horizontal gene transfer from human papillomavirus 18(HPV18) is different from Henrietta Lacks Genome mainly due to the number of chromosomes, but there are also a few other ways. The number of chromosomes varies from cancer formation to its culture due to the cells rapidly dividing. An approximation of the hypotriploid chromosome is (3n+), which is between 75 and 81 chromosomes with 22 to 25 clonally abnormal chromosomes referred to as the HeLa signature chromosomes. They are derived from multiple patent chromosomes, posing a challenge in computing the numbers.

Researchers also realized that the aberrant karyotypes could be Human papillomaviruses (HPVs) integrated into the cellular DNA in cervical cancers. The five HPV 18 integration are 3 on normal chromosome8 at 8q24, and two are located on the derivative chromosomes der(5)t(5;22;8)(q11;q11q13;q24), and der(22)t(8;22)(q24;q13), a consequent chromosome 8q24 material. CGH detected the 8q24 copy increase number. The dual color FISH together with a c-MYC probe revealed co-localization in all HPV18 integration sites, clearly showing a dispersion, and application of c-MYC gene line formed, and viral insertion triggered its likely at a single site. The chromosome structure and number aberration by SKY, genomes which are off balance by CGH, and FISH localization of HPV 18 integration, are representatively used in the advanced stage of cervical cell carcinomas.

The HeLa genome has been stabilized following many years of cultivation, hence gene alteration can be present in the primary tumor, and give a reflection of developmental cervical cancer.

HeLa cells complete genome was sequenced, and published on March 11, 2013 without consent from Lacks' family, which raised concern among them. At the same time the authors purposely withheld access to the sequenced data. Jay Shendure, the sequencing project leader at the University of Washington, brought forth a paper in March 2013 ready for publication, but was not released as the Lacks' family's private concerns had not been thoroughly addressed. On the 7th of August the same year, NIH director Francis Collins, made an announcement on the controlled policy access of the cell genome, a decision that had been arrived at following three meetings with Lacks' family.

The committee set up to review data access requests by researchers in the medical field was to strictly abide by the policies in the HeLa Genome Data Agreement Use. All

NIH-funded researchers were required to deposit all information to a single database for information sharing in the future. The committee comprises of six who are; representatives from bioethics, scientific or medical fields, and two members of Lacks' family. In a recent interview, the NIH director acknowledges Lacks' family's willingness to participate following the situation that was put on them by Henrietta Lacks, who changed how cancer treatment would henceforth take place. He describes the chain of events as 'powerful', and that it has addressed the issues of 'science, scientific history and ethical concerns.

Cell Discoveries and Complications

Due to HeLa cells adaptation to growth in tissue culture plates, they are sometimes difficult to control. Through improper maintenance, they are known to interfere with the cell cultures in the lab. Hence vague biological research and researchers were left with no option but to declare most of their findings as inaccurate. The magnitude of HeLa cells contamination remains unknown because most scientists are ignorant in testing the purity or identity of the already established HeLa cells. A substantial fraction of intro cell lines that show the contamination was in the range of 10% to 20% according to Stanley Gartler in 1967, and Walter Nelson-Rees in 1975 when they released the first publication on the infection of cell lines by HeLa.

Michael Gold, a science writer, in his book "A Conspiracy of Cells" wrote about the HeLa cell contamination problem. In his opinion, Nelson-Rees' work of identification is a pervasive worldwide problem affecting even the laboratories of the best scientists and researchers, such as Jonas Salk and many possibly career-ending efforts to address it. According to him, the HeLa contamination problem almost led to a cold war scenario. President Richard Nixon launched the war against cancer, with the USSR and USA to come together in this only to realize HeLa contaminated the exchanged cells. Gold contends that the HeLa problem was amplified by emotions, egos, and a reluctance to admit mistakes. Nelson-Rees explains that It's all human error. The unwillingness to throw away countless hours of what was regarded as quality work of research, interfering with the grant that was in the process of being applied for, and in haste about who was to come out with the paper first, which is not solely limited to cancer research and biology. Scientists make numerous mistakes, and experience the same problems as well.

Instead of addressing the HeLa cell contamination, scientists and science writers still document it as just an issue of corruption not by human error but by the stubborn, proliferating and overpowering nature of HeLa cell. Recent studies highlight that cross- contamination are an ongoing problem in modern cell cultures according to the International Cell Line Authentication Committee (ICLAC).

Cross contamination and misidentification are still rampant in research, most of which is made from the onset implying all work on these cell lines is incorrect since it may be from a different tissue or species. A cell line is regarded as misidentified if it no longer corresponds to the individual from whom it was first obtained. Cases of misidentification through cross contamination are rampant because of the aggressive growth of cell lines in a culture

Leigh Van Valen describes HeLa as a contemporary creation of a new species, Helacyton gartleri, due to its ability to replicate indefinitely and their many non-human chromosomes. It takes after Stanley M. Gartler, who is credited by the latter for his input in the success of the species. His argument was on the following foundation:

- Incompatibility of chromosomes of HeLa cells with those of human

- Ecological niche of HeLa cells

- HeLa cells ability to expand, and withstand well surpassing the expectation of individual cultivator. For this, HeLa is a species as it has own clonal karyotype.

Van Valen created a new family Helacytidae, and genus Helacyton, as well as a new species of HeLa in the same paper. Contrary, the prominent evolutionary biologists and scientists after him have not taken this seriously. From his work, HeLa does not meet the criteria of an independent unicellular asexually reproducing species, and hence can not be a species in itself. To support this he said, it was because of the HeLa karyotype instability and lack of an ancestral descendant lineage

Family Retribution and Recognition

Henrietta Lacks, together with her family had no consent, to her removal since at the time it was not required nor customarily sought. The cells are used for medical, and commercial purposes. In 1980, family medical records were published without their consent, and the issue was brought in Supreme Court of California. The court ruled that a patient's discarded tissue or cells are not their property, and can be commercialized.

In March 2013, researchers published the DNA code, or genome, of a strain of HeLa cells. The Lacks family discovered this when Rebecca Skloot, an author informed them. There were objections from the Lacks family on the genetic information that was available for public access. Jeri Lacks Whye, Henrietta Lacks' grandchild, informed the New York Times that their concern was privacy. The information being given out to the public about their grandmother, how this information talks about their children, grandchildren and future generations to come. In the same year, a paper on HeLa's cell line genome was submitted for publication by another NIH-funded group which had been working independently.

An agreement was reached between Lacks' family, and the National Institutes of Health that gave the former some power in access to the cells' DNA code, along with a promise for recognition in scientific papers for their valuable contribution. The two Lacks' relatives were to be included in the six-member committee which would regulate access to the code.

Morehouse School of Medicine held its first HeLa Women's Health Conference in 1996. The event was to honor Henrietta Lacks from whom the cell line was named after, and to recognize the significant contribution made by African Americans to medical research and practice, with Dr. Pattilo leading the annual events. The Turner Station is where yearly celebrations take place for Lacks' contribution. Robert Ehrlich, a US congressman from Maryland at such an event, presented a resolution recognizing Lacks' contribution to medical science and research.

Sometime later in 2010, a Clinical and Translation Research came up with the yearly Henrietta Lacks' memorial lecture series at Johns Hopkins. The events were to commemorate Henrietta Lacks, and the global impact HeLa cells has brought in medicine and research.

Similarly, Morgan State University in Baltimore City, Maryland, in 2011, gave Lacks a prestigious honorary doctorate in public service. The Evergreen School District in Vancouver, Washington likewise acknowledged her contribution, and in 2011 named their new high school the Henrietta Lacks Health and Bioscience High School, becoming the first known organization to memorialize her through naming it in her honor publicly.

In 2014, Lacks was inducted into the Maryland Women's Hall of Fame.

Where to Learn More

Written Works

The HeLa cell line first came to light through two popular articles written by Michael Roger in March 1976. These two favorite items are the Rolling Stone and The Detroit Free Press. Later in 1998, Adam Curtis through a BBC documentary film sought to bring out the story of Henrietta Lacks. The film was called The Way of All Flesh.

Hela cell line and Lacks family story continued to be told by different authors. One dedicated author was Rebecca Skloot. Rebecca's work is documented in articles published in 2000, 2001, and her book The Immortal Life of Henrietta Lacks(2010).

The Immortal Life of Henrietta Lacks (2010) by Rebecca Skloot is a nonfiction book. In 2011 the book was top at the National Academies Communication Award for its contribution in helping readers understand topics related to science, engineering and medicine.

Rebeca's book mainly is centered around Henrietta Lacks, and the immortal cell line derived from Lacks's cervical cancer cells in 1951. Notably, the book is known for its deep science writing with a keen focus on ethical issues of class and race in the medical research field. Rebecca Skloot goes on to say that some of the information used in the book were taken from Deborah Lacks, Henrietta Lacks's daughter.

A critical review of this book was mostly favorable. More than 60 media outlets named this book as the best book of the year. Lisa Margonelli, when reviewing this book in The New York Times Book Review, expressed that Skloot was able to break into what most people would refer to as boring scientific facts, and tell the tale of how a family and their mother continued to live despite the disease.

Dwight Garner of The New York Times also saw the beauty of the book, and said that he found it difficult to put the book down. Garner described it as one of the best non-fiction books he has ever read.

While trying to remain positive, one reviewer with The New Atlantis majorly questioned the ethical arguments surrounding tissue markets and informed consent involving scientists such as Chester M. Southam. He claimed to have found factual errors. First of his allegations was one related to the role of HeLa cells in early space missions. The second one was of the claim that if all the cells were collected, and put on a weighing scale that they would amount to about 50 million metric tons. He explained that it was merely a theoretical calculation, and that it wasn't possible to save that many cells, and weigh them. He further added that scientists did the actual calculations of the real figures, and corrected before the book went into publishing.

Some of the awards that the book won include; National Academies Best Book of the Year Award,[7] the American Association for the Advancement of Science's Young Adult Science Book Award,[8] and the Wellcome Trust Book

Prize-, an annual award for an outstanding fiction or non-fiction work on the field of health and medicines. It also bagged The Heartland Prize for non-fiction, a Salon Book Award, and a 100 New York Times Notable Books of the Year.

The book went ahead, and was adopted as a reading guide to more than 125 universities, and was taught in all areas of the classroom from high schools to doctoral classes.

Despite its highs, the book also had its lows. For instance, in September 2015, schools in Knox County, Tennessee faced a demand from a parent to have the book taken down from the library shelves and classrooms for allegations that the nature in which Lacks discovered her tumor in the brain was presented in a "pornographic" manner.

Initially, the book was released in hardcover, and published by Crown on February 2, 2010. Random House Audio released an audio book on the same day. The narrative was done by Casandra Campbell, and Bahni

Turpin (ISBN 978-0-307-71250-9). Broadway Books published a paperback edition on March 8, 2011 (ISBN 978-1-4000-5218-9). So far this book has been translated into more than 25 foreign language editions.

The Movie

The first announcement that Oprah Winfrey and Alan Ball were developing a film project of Skloot's book came in 2010 from HBO. However, the filming commenced way later in 2016. In the movie, the great Winfrey took the leading role of Deborah Lacks, Henrietta's daughter. Though NBC's Law & Order had earlier aired their version of Lacks story in 2010 through a fictionalized story "Immortal," reviews stated it to be "shockingly close to the actual story."

The Immortal Life of Henrietta Lacks by Rebecca Skloot is a drama television film directed by George C. Wolfe, and features Oprah Winfrey. The majority of the film is based on Rebecca Skloot's book of the same name, and is the story of Henrietta Lacks, the diagnosis of cervical cancer in 1950, and how later her cancer cells would then come to impact the course of cancer treatment today.

The official announcement of the movie was on May 2, 2016. George C would be the scriptwriter and director. Wolfe, with Oprah Winfrey who was the producer, and Lacks' daughter Deborah. Rose Byrne acted as Rebecca Skloot, the author of the book about Henrietta. She befriends Deborah while reporting on her mother's life. Renée Elise Goldsberry was cast as the so-called Henrietta Lacks. The main supporting was completed in August 2016. Courtney B. Vance, Ruben Santiago-Hudson, Reg E. Cathey and Leslie Uggams were among them. September 21, 2016, is when The Immortal Life of Henrietta Lacks started filming in Baltimore after a couple of shootings in Atlanta.

The film has been a passion project for the TV mogul and Oscar-nominated actress,Oprah, who teamed up six years ago with Six Feet under and True Blood creator Alan Ball to produce the feature length adaptation of Rebecca Skloot's book.

George C. Wolfe who is a known Broadway director of shows such as HBO's Lackawanna Blues is the writer of the movie, and will also double as the director when the shooting starts in the summer. Oprah has had great acting experience in supporting roles for shows such as the Butler and The Color Purple.

Oprah and Oscar winner Ball will be the executive producers with help from Peter Macdissi, who is the executive producer of Banshee. Carla Gardini, executive producer of The Hundred-Foot Journey, and Lydia Dean Pilcher who did HBO's You Don't Know Jack. Skloot serves as co-executive producer, while Henrietta Lacks' sons David Lacks Jr., Zakariyya Rahman and granddaughter Jeri Lacks are consultants.

2013 for the first time saw members of the Lacks family author their stories. The oldest son of Lack and his wife wrote a short recap of their lives called "Hela Family Stories: Lawrence and Bobbette". These stories were firsthand accounts of their memories of Henrietta Lacks while she was alive. Also depicted in the stories were the struggles, and their efforts to keep the younger children safe after the demise of their mother.

The Book

The Immortal Life of Henrietta Lacks took Skloot more than ten years to research, and write. However, the book hit the bestseller list of The New York Times, and remained at that position for more than four years.

www.ingramcontent.com/pod-product-compliance
Lightning Source LLC
Chambersburg PA
CBHW030102230526
45471CB00003B/1208